Marcela Liege da Silva
Eliel Farias
Edvan Chagas

Use of the Biospeckle technique on Eugenia stipitata seeds

Marcela Liege da Silva
Eliel Farias
Edvan Chagas

Use of the Biospeckle technique on Eugenia stipitata seeds

Biometric characterisation, germination and viability of araçá-boi seeds

ScienciaScripts

Imprint

Cover image: www.ingimage.com

This book is a translation from the original published under ISBN 978-613-9-66940-0.

Publisher:
Sciencia Scripts
is a trademark of
Dodo Books Indian Ocean Ltd. and OmniScriptum S.R.L publishing group

120 High Road, East Finchley, London, N2 9ED, United Kingdom
Str. Armeneasca 28/1, office 1, Chisinau MD-2012, Republic of Moldova, Europe
Printed at: see last page
ISBN: 978-620-8-11152-6

DEDICATORY

To my parents Jânio Antônio da Silva and Tânia Fátima Silva for their unconditional love, support for my studies and for shaping my character.

To my siblings Priscila Elisama and Joeser Elias for their friendship and unity.

Dedication.

ACKNOWLEDGEMENTS

I thank God for the life and wisdom he has given me.

To the Postgraduate Programme in Biodiversity and Biotechnology of the Bionorte Network for the opportunity to take the course.

To the National Council for Research and Technological Scientific Development for the scholarship awarded, which allowed for excellent working conditions.

To the Brazilian Agricultural Research Corporation for the availability of physical space and support in conducting the experiments.

To my advisor and friend D.Sc. Edvan Alves Chagas for his credibility in my work and for the opportunity to grow professionally.

To my friend and researcher M.Sc. Rosemary Vilaça, for her guidance in the work carried out, her concern for the results, and for the friendship we gained.

To researcher Oscar José Smirdele, for his guidance and participation in the work, and especially for his attention and help in writing the thesis.

To Professor D.Sc. Eliel Eutério Farias for his guidance and encouragement in carrying out a totally innovative and promising piece of work.

To my great friend Pollyana Cardoso Chagas for almost twenty-five years of friendship, complicity, joy and, above all, for being part of this achievement.

To my dear friends Pedro Henrique and Raissa, for the games, the mess and the moments of relaxation at the weekends, good memories.

To my fellow Embrapa members and colleagues, for their help and companionship from collecting the fruit in the field to carrying out the experiments, without naming names so as not to be unfair to anyone.

Thank you all.

A friend isn't the one who helps you when you're unwell.

That's easy. Difficult is the friend who loves to see you shine.

Fernanda Estellita

SUMMARY

The Amazon has countless plant species with great economic potential. Among these species, fruit trees have stood out for generating income and food for the population. These species are still understudied and their main obstacle is the lack of knowledge about their propagation, such as Eugenia stipitata, whose main problem is the dormancy of its seeds. The aim of this study was to carry out biometric characterisation, germination and a study of conventional and modern techniques to assess how to overcome the dormancy of araçá-boi seeds. Three experiments were carried out in order to: 1 - carry out biometric characterisation and obtain the water absorption curve and vigour study of araçá-boi seeds; 2 - evaluate the efficiency of osmoconditioning promoted by potassium nitrate and sucrose on the germination of araçá-boi seeds, at different immersion times; and 3 - study the use of the biospeckle technique to determine the viability and vigour of the seeds, correlating it with the germination behaviour and development of the seedlings in the field and checking the possibility of using the technique as an auxiliary tool to reduce the time taken to evaluate the treatment to overcome dormancy in araçá-boi seeds. The plant population established in Manaus-AM produces small, medium and large seeds weighing 0.08, 064 and 1.37 g and with 70.2%, 68.2% and 47.5% humidity, respectively. The most water was absorbed by the seeds in the first hour of absorption, regardless of size, and the speed of water absorption was not influenced by size. The soaking curves characterised over 144 hours, regardless of seed size, showed two physiological phases. Large and medium-sized seeds showed lower conductivity values, obtaining higher germination percentages because they had higher vigour indices. Osmotic conditioning resulted in 21% of germinated seeds in 14 days of cultivation. 100% of seeds germinated when osmoconditioned in KNO3 for 12 hours, up to 70 days of cultivation. The use of the biospeckle technique was efficient in determining the viability and vigour of araçá-boi seeds. Large and medium-sized seeds showed higher activities. The germination of large and medium-sized seeds and the development of the seedlings correlated with the activity observed by the biospeckle technique. The initial results show that it is possible to carry out further studies in order to obtain an efficient methodology for evaluating treatments to overcome dormancy in araçá boi seeds.

Keywords: Eugeniastipitata ; Native Fruit Trees; Osmoconditioning; Recalcitrants;

SUMMARY

INTRODUCTION	**6**
CHAPTER 1	**10**
CHAPTER 2	**21**
CHAPTER 3	**28**

INTRODUCTION

The Amazon has one of the highest levels of biodiversity in the world and there are countless plant species with great economic potential. According to Clement et al. (2007), fruit growing represents an important segment of agriculture for the development of the Amazon region and is included among the environmentally sustainable and economically viable options, due to the potential for using the region's native genetic resources and the number of labourers involved.

Among the predominant species in the Amazon region, the Myrtaceae family is of great economic importance due to its edible fruits, such as Psidium guajava (guava), Myrciaria cauliflora (jabuticaba) and Eugenia uniflora (pitanga). However, others have been standing out, such as Eugenia stipitata ssp. Sororia Mc Vaugh (araçá-boi), with great potential for agro-industrial use, as they have good physical-chemical characteristics and sensory attributes that are acceptable (SOUZA FILHO et al., 2002; ROGEZ et al., 2004).

Most of the fruit species native to the Amazon are undomesticated and little studied, with the biggest obstacle being the lack of knowledge about their propagation. In this way, mastering technologies and establishing a cultivation system for these fruit trees is essential if efficient production is to be possible, on a commercial scale and in a rational and sustainable manner.

Another important factor is that most native species produce recalcitrant seeds with some degree of dormancy, which makes it difficult for producers to handle them and produce seedlings. Seed dormancy plays an important role in the perpetuation of species, but it is a major problem for commercial propagation, especially if the dormancy is inherent to the seed.

The intrinsic factors related to seed dormancy are mainly the hardness and impermeability of the tegument to water and gases, the immature embryo and the presence of germination inhibitors (BEWLEY and BLACK, 1994; BASKIN and BASKIN, 1998). Physiological dormancy involves various mechanisms located in the embryo, tissues and adjacent structures such as the tegument and endosperm.

In view of the above, knowledge of pre-germination treatments that guarantee efficient seed germination is necessary in order to achieve good quality propagation and seedling production practices for araçá-boi.

In this context, the aim was to carry out biometric characterisation, germination and a study of conventional and modern techniques for overcoming the dormancy of araçá-boi seeds. To this end, the thesis was formatted into chapters for a better understanding of the study. After formatting, the chapters were transformed into articles and prepared in accordance with the rules of each journal to be submitted.

Chapter I presents "BIOMETRIC CHARACTERISATION, WATER ABSORPTION CURVE AND VIGOUR IN ARAÇÁ-BOI SEEDS".

Chapter II presents "EFFICIENCY OF CONVENTIONAL TECHNIQUES AND OSMOCONDITIONING IN THE GERMINATION OF ARAÇÁ-BOI SEEDS".

Chapter III presents *BBIOPECCKEE* AS AN AUXILIARY TOOL IN THE EVALUATION OF DORM OVERVIEW IN ARAÇÁ- BOI SEEDS".

LITERATURE REVIEW

The araçá-boi is a fruit species native to the Peruvian Amazon and the western region of the Brazilian Amazon and is usually grown on a small scale in Peru, Bolivia, Ecuador, Colombia and Brazil (PINEDO et al., 1981; CLEMENT, 1989). It arouses interest because of the organoleptic qualities of the fruit and the plant's production rate. Its juicy fruit has a pleasant aroma and flavour and can be eaten as a soft drink, jam, nectar, jelly and liqueur (PICON, 1989; ANDRADE et al., 1997).

The araçá-boi plant is a shrub up to 2.5 metres high, densely branched and leafy, with elliptical, dark green leaves (MENDES, 2011). The fruit is a berry with an average weight of 200 g. When ripe, it has a very pleasant aroma. The pericarp is canary yellow, velvety and thin; the flesh is juicy with a pleasant, acidic flavour (MCVAUGH, 1956).

The plant produces practically all year round (MENDONÇA et al., 2001) and with this, depending on the management of the plantation, there is the possibility of generating work and income continuously in the field, without the problem of seasonality, which is common in many crops.

The seed coat is thin, leathery and thick. The outer surface is dark brown, with a leathery appearance. Inside, the seed coat is smooth and light brown. As for shape, they can be reniform, flattened, discoidal, longer than they are wide or vice versa, almost spherical, with the predominant morphology being reniform with lateral flattening (ANJOS and FERRAZ, 1999).

The seeds of E. stipitata are dormant and according to Pinedo et al. (1981), when dried under shaded conditions, they begin to lose viability after 24 hours, showing a germination rate of 70% after 5 days. The same author attributes this dormancy to the tegument, which when dried becomes resistant to rupture by the embryo.

In view of the above, knowledge of pre-germination treatments that guarantee efficient seed germination is necessary in order to achieve good quality propagation and seedling production practices for araçá-boi.

Most work on overcoming seed dormancy is centred on physical (scarification, removal of the endocarp) and chemical (sulphuric acid) treatments. In Eugenia stipitata, the most efficient method for overcoming dormancy consists of removing the impediment at the point where the primary root protrudes (MENDES, 2011), however, this technique can damage the embryo, causing the seedling to develop abnormally.

New ways are being studied to overcome seed dormancy without jeopardising plant formation. Osmotic conditioning methods have been tested, which consists of controlled hydration of the seeds in an osmotic solution, stimulating the seed's metabolism without, however, allowing the primary root to emerge (NASCIMENTO, 2004).

Osmotic conditioning favours germination due to the accumulation of solutes from the start of the seed's metabolism, promoting radicle emergence and seedling formation in a shorter space of time (Albuquerque et al., 2009). Osmotic conditioning uses inorganic osmotic agents such as sodium chloride (NaCl), potassium nitrate (KNO3), magnesium sulphate (MgSO4) and organic osmotic agents such as mannitol, glycerol, sucrose and polyethylene glycol (PEG) (KISSMANN et al., 2011).

The need to speed up the germination process of seeds and rapid and efficient techniques for assessing their viability has led to the use of non-destructive methods, since they preserve the product being analysed.

The speckle correlation technique has proved to be an excellent alternative for studying biological activity to reveal random intercellular movement, as well as being a non-destructive, real-time test. This movement, caused by metabolic activity, can be greater or lesser depending on the state of maturation, growth or deterioration of the biological material, as well as the amount of water, temperature and lighting (RODRIGUES, 2003).

The speckle effect is a phenomenon of interference between mutually coherent waves with a random variation in phase. The result of this type of interference is a distribution of light intensity that is stationary in time but random in space (PIRES, 2003).

Dynamic speckle, when it comes from biological material, is also known as biospeckle. This term is defined as the interference figure formed when a laser is shone on a dynamic process and the so-called boiling phenomenon is observed. This dynamic process can be a painting being dried (SILVA, 2007), the transmission of heat in a plate (BRAGA JÚNIOR, 2000) or the activity of a living organism (RODRIGUES, 2003).

OBJECTIVES

3.1 GENERAL OBJECTIVE

Biometric characterisation, germination and study of conventional and modern techniques for overcoming dormancy in araçá-boi seeds.

3.2 SPECIFIC OBJECTIVES

To carry out biometric characterisation, obtain the water absorption curve and study the vigour of araçá-boi seeds.

To evaluate the efficiency of sodium hypochlorite in the scarification process and osmoconditioning promoted by potassium nitrate and sucrose on the germination of araçá-boi seeds, at different immersion times.

To study the use of the biospeckle technique to determine the viability and vigour of seeds and correlate this with the germination behaviour and development of seedlings in the field.

To study the possibility of using the biospeckle technique as an auxiliary tool to reduce the time needed to evaluate the treatment to overcome dormancy in araçá boi seeds.

CHAPTER 1

BIOMETRIC CHARACTERISATION, WATER ABSORPTION CURVE AND VIGOUR IN ARAÇÁ-BOI SEEDS

Summary - The araçá-boi (Eugenia stipitata) is a fruit that has aroused great interest because it has quality pulp and a high rate of plant production. Its seeds are recalcitrant and have a relatively short viability period. The absorption of water by seeds is a mechanism for resuming embryo growth, with subsequent breaking of the tegument by the radicle and is directly associated with seed viability. The electrical conductivity of seeds is a vigour test in which the electrical current from the loss of leachate from the inside of the cell to the outside medium is assessed. This process is related to membrane permeability, with lower values indicating greater vigour. Therefore, studying the water absorption curve of araçá-boi seeds will help to characterise their germination process. In this context, studies were carried out to characterise the mass, biometrics, water content, electrical conductivity and water soaking of the seeds. It was observed that the plant population established in Manaus-AM produces small, medium and large seeds with 70.2%, 68.2% and 47.5% humidity. The greatest increase in water occurs in the first hour of absorption, regardless of seed size, and the speed of water absorption is not influenced by seed size. The soaking curves, characterised over a period of 144 hours, regardless of seed size, show changes in two physiological phases. Large and medium-sized seeds have lower conductivity values and higher germination percentages because they have higher vigour indices.

Keywords: Viability. Recalcitrant. Seeds;

INTRODUCTION

The araçá-boi (Eugenia stipitata ssp. sororia McVaugh) is a promising alternative crop for the Amazon countries and, although it is not very widespread, it is attracting interest due to the good physical and chemical characteristics of its fruit and the sensory attributes that are well accepted (SOUZA FILHO et al., 2002; ROGEZ et al., 2004).

Despite its ease of cultivation, the araçá-boi has physiological characteristics that require care in the seedling formation process, as they have recalcitrant seeds, i.e. they cannot tolerate dehydration and low storage temperatures and their viability period is relatively short.

The reserve substances and other compounds present in the seed are fundamental for the propagation of native species, as they are related to germination, storage potential and seedling establishment (MENDES, 2011).

According to Carvalho and Nakagawa (2000), in general, larger seeds are better nourished during their development, have a well-formed embryo and a greater quantity of reserve substances and are therefore the most vigorous. The greater amount of reserves increases the likelihood of success in establishing the seedling, as it allows it to survive for longer in unfavourable environmental conditions.

Popinigis (1985) states that seed size in many species is indicative of physiological quality, so biometric characterisation of fruit and seeds is important for taxonomy, identifying varieties and verifying the occurrence of phenotypic variations (CARDOSO and LOMÔNACO, 2003; PINTO et al., 2003).

The process of seed germination begins after the absorption of water, with the subsequent breaking of the tegument by the radicle and the subsequent development of the main root. This absorption follows a three-phase pattern in most species, where phase I occurs rapidly due to the difference in water potential between the seed and the substrate (MARCOS FILHO, 2005). Phase II of germination is characterised by a drastic reduction in the speed of absorption, with an increase in the diffusion of solutes to regions of marked metabolism, mainly in the region of the embryo. Phase III begins with the emission of the primary root and only occurs in non-dormant seeds. These three phases give rise to the seed water absorption curve, contributing to the standardisation of tests to assess the physiological quality of seeds (FERREIRA et al., 2006; CARVALHO and NAKAGAWA, 2012).

There are also other important evaluations for understanding germination processes. The electrical conductivity of seeds is considered to be a test of vigour, in which the electrical current is evaluated through a conductivity bridge in the soaking solution and the loss of leachates from the inside of the cell to the outside environment, involving biochemical processes entirely related to the disorganisation of cell membrane systems (MARQUES et al., 2002). In the electrical conductivity test of seeds, the lowest values correspond to the lowest release of exudates, indicating greater vigour. In this context, the aim was to carry out biometric characterisation, obtain the water absorption curve and study the vigour of araçá-boi seeds.

MATERIAL AND METHODS

Material collection

The seeds used in this study came from the fruits of various populations of E. stipitata at different stages of ripeness, collected in the city of Manaus between 7 and 9 November 2013. The seeds were extracted manually, followed by washing in running water using a sieve to completely remove the mucilage. They were then left to dry for 2 hours at an average temperature of 25°C. The seeds were then quantified and separated into three sizes manually, visually, followed by individual weighing of the seeds to confirm this separation into three groups: 1- Large, 2 - Medium and 3 - Small. This resulted in seeds with masses of more than 0.78g in the large group, between 0.77g and 0.19g in the medium group and less than 0.18g in the small group.

Seed biometrics and material weighing

150 seeds were selected from each group and left in a 25° C environment for 12 hours, after which their mass (g) was obtained using a 0.001 g precision balance. To determine the biometry, the length, width and thickness (mm) of the seeds were measured using a digital caliper.

Water content (%)

Twenty seeds from each group were used and placed in plastic cups, with four replicates of five seeds. The seeds were soaked in water for 72 hours and weighed on a 0.001 g precision scale, after which they were placed in an oven at 105°C for 24 hours (Brasil, 2009). The water content of the seeds was calculated as follows:

$$\text{Water content (\%)} = 100 \times (P1-P2)/P1$$

Where: P1 = mass of the wet sample; P2 = mass of the dry sample.

Electrical conductivity

The leaching pattern of the exudates was assessed for the seeds from the three groups, using four replicates of 25 seeds, previously weighed on an analytical balance with a precision of 0.001 g and placed in containers containing 75 mL of distilled water, kept at room temperature of 25°C for 24 hours. After this period, the immersion solution was gently stirred and the conductivity reading was taken on a bench conductivity meter, with the results divided by the initial mass of the seeds. The values were expressed in LIS cm^{-1} g^{-1} of seed.

Soaking

100 seeds chosen at random from the three groups were used and placed in plastic cups with four replicates of 25 seeds. The initial mass of each repetition was determined and 75 mL of water was added to each cup. The seeds were soaked in water for periods of 01, 02, 03, 05, 09, 15, 24, 48, 72, 96, 120 and 144 hours, at an ambient temperature of 24°C + 2°C.

At each time interval, initially shorter (1 hour) and then longer (24 hours), successive weighings were made to quantify the wet mass gain that occurred in each period and over time. After each period, the seeds were removed from the water and dried on paper towels, weighed and returned to the plastic containers. Dull seeds were considered to be dry because of the film of moisture on the outside of the tegument.

Seed soaking was measured by wet mass gain in successive weighings up to 144 hours.

The data obtained was analysed using the computer program Sisvar (FERREIRA, 2011). The values were subjected to analysis of variance and their means were compared using the Tukey test at 5% probability (GOMES, 2000).

RESULTS AND DISCUSSION

The results obtained from determining the individual mass of 150 seeds showed that the visual separation of the initial sample into three groups made up of large, medium and small seeds was effective (Table 1). An average of 1.37g was obtained for large seeds, 0.64g for medium seeds and 0.08g for small seeds.

Table 1 - Average values* of length, width, diameter and mass of the Araçá-boi seeds used in this work, Boa Vista-RR, 2015

Table 1- Mean values * of length, width, diameter and mass of the Araca-boi seeds used in this study, Boa Vista-RR, 2015

Size of Seeds	Length (mm)	Width (mm)	Diameter (mm)	Mass (g)
Small	7,17	6,09	4,90	0,08
Averages	14,27	10,55	8,06	0,64
Large	19,68	13,04	9,67	1,37

*Averages of 25 seeds.

*Average of 25 seeds.

Statistical analysis of the water absorption results (Table 2) revealed a non-significant interaction between the sources of variation, seed size and immersion periods. Table 2 - Mean square values and significance for water absorption data for 144 hours in small, medium and large araçá-boi seeds, Boa Vista, RR, 2015

Table 2 - Values of mean squares and their significance for the absorption water for 24 hour of large, medium and small araçá-boi seed, Boa Vista, RR, 2015

Sources of Variation	Degrees of Freedom	Mean Squares
Soaking period (PE)	12	1,38**
Seed size (TS)	2	2179,75**
Size x Period (PExTS)	24	0,16$^{n.s}$
Error	117	0,41
Total	155	
C.V.%		8,5

$^{ns,\ **}$, not significant and significant at 1% by Tukey's test, respectively. C.V.%= percentage coefficient of variation.

$^{ns,\ **not}$ significant and significant in 1% in Tukey test, respectively. C.V.%= percent coefficient of variation.

Figure 1 shows the wet mass of the seed, thus verifying the evolution of the processes through the water absorption curves of the small, medium and large seeds over the observed period.

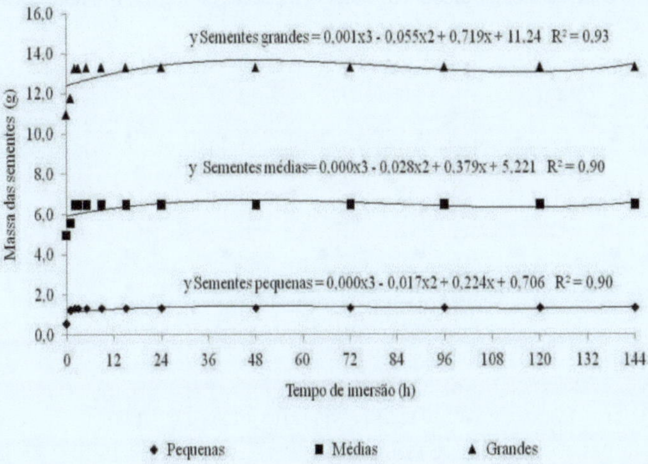

Figure 1 - Mass (g) of araçá-boi seeds (small, medium and large) soaked in water for 144 hours.

Figure 1 - Mass (g) of seeds (small, medium and large) araçá-boi soaked in water until the period of 144 hours.

Figure 2 shows the approximation of the wet mass curve of large (G), medium (M) and small (P) araçá-boi seeds in two physiological phases of soaking.

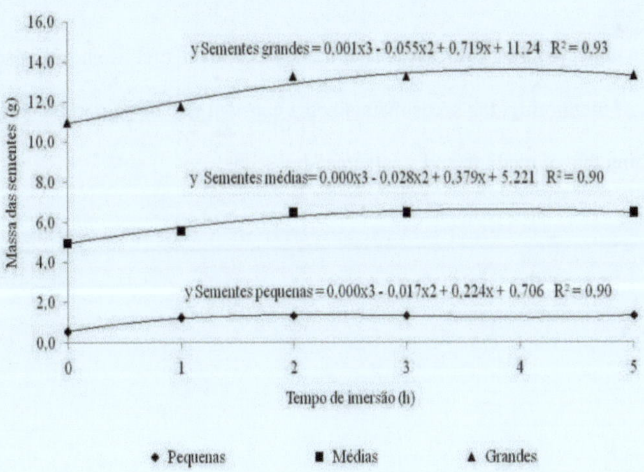

Figure 2 - Mass (g) of araçá-boi seeds (small, medium and large) soaked in water for 5 hours.

Figure 2 - Mass (g) of seed (small, medium and large) of Araçá-boi soaked in water until the period of 5 hours.

Figure 3 shows the increase in the amount of water in the seeds during the 144 hours of soaking.

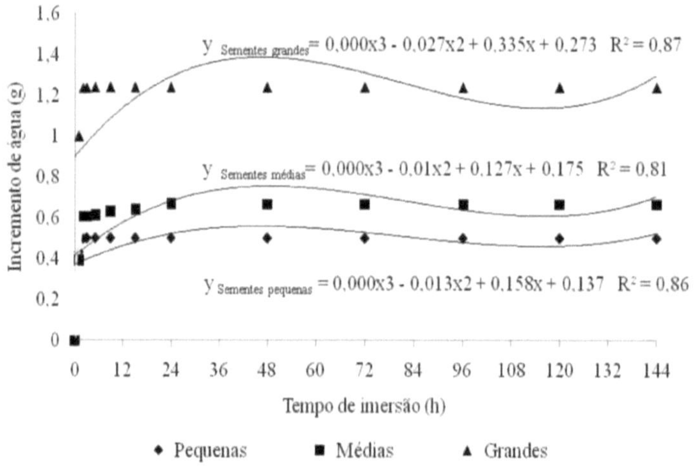

Figure 3 - Water increment (g) found in small, medium and large araçá-boi seeds immersed in water for 24 hours.

Figure 3 - Water Increment (g) found in the seeds (small, medium and large) of Araçá- boi immersed in water up to the period of 144 hours.

Figure 4 shows an increase in the amount of water in the seeds in the first two hours of soaking. In the following hours of monitoring, the seeds maintained a constant rate of gain in the amount of water (mass), approaching that of medium and small araçá-boi seeds.

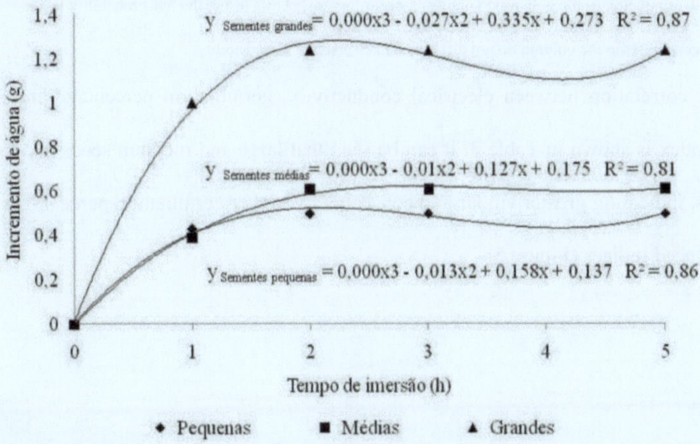

Figure 4 - Water increment (g) found in small, medium and large araçá-boi seeds immersed in water for up to 5 hours.

Figure 4 - Water Increment (g) found in the seeds (small, medium and large) Araçá-boi immersed in water until the period of 5 hours.

The average electrical conductivity values obtained for small, medium and large araçá-boi seeds are shown in Table 3. A relationship was observed between seed size and electrical conductivity. Small seeds had higher average conductivities, which indicates lower vigour due to the greater release of electrolytes into the immersion solution.

Table 3 - Average values of electrical conductivity (EC) of seeds immersed in water for 24 hours, germination percentage and seedling emergence rate obtained from araçá-boi seeds after 100 days.

Table 3 - Mean values of electrical conductivity (EC) of immersed seeds in water within 24 hours germination and seedling emergence speed obtained from Araçá-boi seeds at the end of 100 days.

Size of Seed	EC (μS cm^{-1} g)$^{-1}$	Germination (%)	Speed emergency (Index)
Large	0,79a	71a	1,94
Averages	1,70a	64a	1,45
Small	12,04b	2 b	0,06

Averages followed by the same letter in the column do not differ according to Tukey's test at the 5% probability level.

Means followed by the same letter in the column do not differ by Tukey test at 5% probability.

The positive correlation between electrical conductivity, germination percentage and seed emergence speed index is shown in Table 3. It can be seen that large and medium seeds had lower conductivity values, indicating greater vigour and consequently higher germination percentages and higher emergence speed indices (Figure 5).

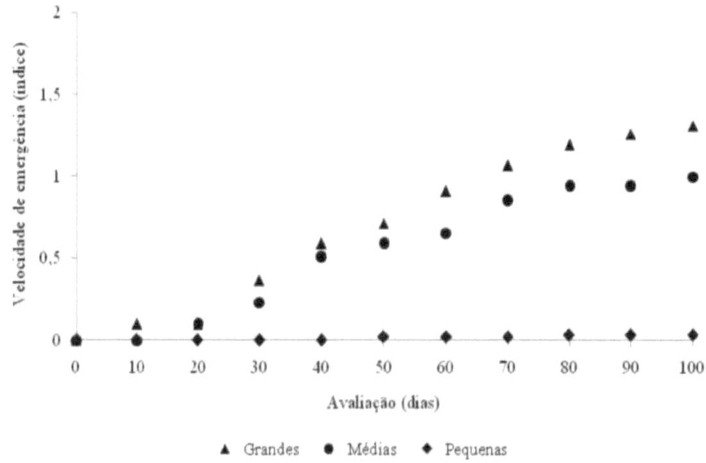

Figure 5-Accumulative seedling emergence speed (index) obtained from araçá-boi seeds up to 100 days.

Figure 5 - Emergency-Velocity (rate) of cumulative seedlings obtained from seeds of Araçá-boi up to 100 days.

The results obtained in this study indicate that the different seed sizes observed are related to the degree of maturity of the seeds, which can be proven by the high water content and electrical conductivity of the small seeds, causing reduced germination. However, further studies could be carried out to better understand this aspect in seeds from other populations.

These values for large and medium seeds were higher than those obtained by Anjos (1999), who studied the morphology, germination and water content of Eugenia stiptata seeds and found an average weight of 0.49g. Thus, the seeds obtained from the plant population established in Manaus-AM were separated into three distinct groups, called small, medium and large. As for the results of the average water content of the seeds, the average humidity values determined in an oven at 105°C for 24 hours were 47.5% for large seeds, 68.2% for medium seeds and 70.2% for small seeds.

According to Carvalho and Nakagawa (2012), cotyledon seeds end phase I with water contents of between 35 and 40 per cent, after which phase II begins. However, each phase can be lengthened or shortened depending on the inherent properties of the seeds and environmental conditions.

Water is a factor that has a decisive influence on the seed germination process. The rehydration of the tissues results from the absorption of water by the seeds and the intensification of respiration and other metabolic activities, which provide energy and nutrients for the embryonic axis to resume its growth. By absorbing water, the seed's cotyledons increase in volume, which causes the tegument to break,

facilitating the emergence of the hypocotyl root axis and other internal structures of the seed (BORGEset al., 2009). Therefore, once the soaking process has begun, this will result in seed germination (MARCOS FILHO, 2005; ANDRADE et al., 2006), continuing the life cycle by distributing plants in nature (SOUZA et al., 2007).

In the seeds, regardless of class, phase I occurred in the first two hours, when there was a rapid increase in wet mass; from this period onwards, water absorption stabilised, which characterises phase II (stationary) of the curve.Costa et al. (2001), in studies of the germination of two types of umezeiro (sour and sweet) as a function of pre-soaking the endocarps in water (0h, 48h, 96h and 144h), found that there was no effect of this treatment on seed germination.

According to Castro and Hilhorst (2004), soaking is a purely physical process, related to the properties of colloids, and depends on water binding to the seed matrix. Cardoso (2004) emphasises that the matrix component is mainly responsible for the movement of water at the start of soaking, but with greater availability of free water the osmotic component becomes very active.

The process of the seed soaking up water triggers sequential metabolic changes that culminate in the protrusion of the radicle when it comes to viable, non-dormant seeds (CARVALHO and NAKAGAWA, 2012). As the seeds evaluated in this study were dormant, they did not show radicle protrusion at the end of the successive weighings carried out over 144 hours of emersion, so phase III was not observed.

Lima et al. (2006) found low germination of freshly collected Caesalpiniaferrea seeds, attributed to low water gain during soaking, showing that the seeds had integumentary dormancy. A similar result was found in this study for large, medium and small araçá-boi seeds.

The speed of water absorption was not influenced by the size of the seeds, indicating that there was no difference in the permeability of the seed coat of the large seeds compared to the medium and small ones used in this study.

The soaking curves, characterised over a period of 144 hours, regardless of seed size, showed the changes in the first two physiological phases. The first phase lasted two hours, and the second phase extended until the end of the 144-hour immersion monitoring. In this phase, the soaking of the seeds remained constant, showing an important advance in the standardisation of vigour tests.

In a study on Coffea arabica L., Malta et al. (2005) also found that small seeds had higher electrical conductivity values. This may have been due to the fact that small seeds have a larger specific area than large ones, which allows them to absorb water more quickly, releasing more electrolytes into the reading solution.

CONCLUSIONS

The plant population established in Manaus-AM produces small, medium and large seeds with 70.2%, 68.2% and 47.5% moisture content.

The greatest increase in water absorption by seeds occurs in the first hour of absorption, regardless of size, and the speed of water absorption is not influenced by size.

The soaking curves characterised during the period evaluated, regardless of seed size, show two physiological phases: large and medium seeds have lower conductivity values and higher germination percentages because they have higher vigour indices.

CHAPTER 2

OSMOCONDITIONING IN THE GERMINATION OF ARAÇÁ-BOI SEEDS

Abstract - The araçá-boi is a fruit native to the Amazon region and its fruit has great potential for agro-industrial utilisation due to the intrinsic characteristics of its pulp. Its seeds are dormant and the main cause may be due to the presence of chemical substances that inhibit germination and may also be caused by the impermeability of the tegument. In this context, the aim of this study was to evaluate the efficiency of osmoconditioning promoted by sucrose ($C_{12}H_{22}O_{11}$) and potassium nitrate (KNO_3) on the germination of araçá-boi seeds, at different immersion times. To this end, the seeds were osmoconditioned in solutions of $C_{12}H_{22}O_{11}$, KNO_3, KNO_3,+ $C_{12}H_{22}O_{11}$ and H_2O for periods of 6, 12 and 24 hours. Germination percentage and germination speed (index) were assessed over 70 days. The design used was entirely randomised in a 4x3 factorial, with 4 replicates of 10 seeds. The osmoconditioning of araçá-boi seeds in a KNO3 solution is effective in promoting germination and germination speed. Araçá boi seeds osmoconditioned in KNO3 have a germination percentage of 99% up to 70 days. The period of osmoconditioning does not influence the germination of araçá-boi seeds.

Keywords - Eugenia stipitata. Dormancy. Priming.

INTRODUCTION

The araçá-boi (Eugenia stipitata McVaugh) is a fruit tree native to the Amazon, belonging to the Myrtaceae family and is widely distributed in the Western Amazon, usually cultivated on a small scale in Peru, Bolivia, Ecuador, Colombia and Brazil (CHÁVES FLORES and CLEMENT, 1984).

The fruit of the araçá-boi tree has great potential for agro-industrial utilisation, due to its good physical-chemical characteristics and sensory attributes that are highly acceptable (ROGEZ et al., 2004).

The germination of araçá-boi seeds is hypogeous, cryptocotyledonous and generally takes between 50 and 280 days, from the beginning to the end of the germination process (ANJOS and FERRAZ, 1999). Some studies point to the seed's inherent dormancy as one of the main causes of this long germination period, as well as the slow emergence of seedlings of this species.

The dormancy of araçá-boi seeds may be due to the presence of chemical substances that inhibit germination, such as abscisic acid and phenolic compounds (Sert et al., 2009). These same authors suggest leaching as a method for removing the dormancy caused by these inhibitors.

The dormancy of araçá-boi seeds can also be caused by the impermeability of the tegument (PINEDO et al., 1981). Gentil and Ferreira (1999) confirmed that the tegument of araçá-boi seeds mechanically restricts the expansion of the embryo. In a study carried out by Mendes (2011), it was revealed that the partial removal of the tegument on the side of the root protrusion favoured the germination of araçá-boi seeds.

Other methods of overcoming seed dormancy can be used, such as osmotic conditioning in organic and inorganic substances, with the aim of forming a greater number of viable seedlings, especially with uniformity.

Osmoconditioning consists of the controlled hydration of seeds in an osmotic solution, stimulating their metabolism without, however, allowing the primary root to emerge (NASCIMENTO, 2004). This conditioning activates the degradation, translocation and assimilation of reserves, allowing the seeds to reach a relatively uniform metabolic state when the water supply is interrupted.

Osmotic conditioning uses inorganic osmotic agents such as potassium nitrate (KNO_3) and organic osmotic agents such as sucrose ($C_{12}H_{22}O_{11}$), water (H_2O), among others (KISSMANN et al., 2011). Various studies have shown that osmotic conditioning accelerates germination, allowing seedlings to emerge more quickly and uniformly in the field (NASCIMENTO; ARAGÃO, 2004, PEREIRA et al., 2008; ARAÚJO et al., 2011).

Germination speed may not be influenced by pre-soaking, which is a treatment that does not necessarily influence the germination of all types of seeds (NASCIMENTO, 2004). According to Nascimento and Aragão (2004) and Pereira et al. (2008) osmoconditioning is more efficient for seeds of low to medium vigour.

The main characteristic of the araçá-boi seeds used in this work is their long germination period, which can take up to 8 months to reach 100% germination, thus requiring treatments to anticipate and standardise germination and increase germination speed in order to obtain seedlings. In this context, the aim was to assess the efficiency of osmoconditioning promoted by sucrose ($C_{12}H_{22}O_{11}$) and potassium nitrate (KNO_3) on the germination of araçá- boi seeds at different immersion times.

MATERIAL AND METHODS

The seeds used in this study were obtained from mature fruits of E. stipitata. The fruit was collected in the city of Manaus between 7 and 9 November 2013. The seeds were extracted manually, followed by washing in running water using a sieve to completely remove the mucilage. The seeds were then left to dry in a natural environment for two hours at an average temperature of 25 °C, after which the medium seeds were selected for the experiment. Average seeds were those with a mass of 64 g.

After obtaining the seeds, they were stored in containers immersed in distilled water, which was changed every three days until the experiment was finalised. The experimental design adopted was entirely randomised, with four replicates of 10 seeds, in a 4 x 3 factorial scheme, with four osmoconditioning agents and three conditioning times as treatments. The first factor consisted of four osmoconditioning treatments: sucrose solution (30 g L^{-1}); KNO3 solution (4.5 g L^{-1} with an osmotic potential of -1.1 MPa and a concentration based on the Van't Hoff equation (HILLEL, 1971)); sucrose solution + KN03; and distilled H_2O alone. The second factor was three conditioning times (6, 12 and 24 hours).

The seeds were removed from the container and then immersed in a 2% active chlorine NaClO solution for 60 minutes to disinfect them. After sanitising, the seeds were washed in distilled water and placed in plastic boxes (Gerbox®) previously lined with filter paper moistened 2.5 times the weight of the paper with the osmoconditioning solutions, according to each treatment. The plastic boxes (Gerbox®) containing the seeds were then kept in a B.O.D (Biochemical Oxygen Demand) chamber at 10 °C, in the absence of light, wrapped in a plastic bag to prevent the solution from evaporating, for the established periods.

After each conditioning period (6, 12 and 24 hours), the seeds were washed in distilled water and placed back in plastic boxes (Gerbox®) to quantify the germination percentage. The plastic boxes were previously lined with filter paper and distilled water was added until a third of the seed was covered. They were then incubated in a B.O.D. at 25 °C for 70 days. At each 7-day interval, the germination percentage and germination speed index (GVI) were determined according to Maguire (1962), where GVI = N1/D1 + N2/D2 + Nn/Dn where N1, N2 and Nn = number of seeds germinated in the first, second and last counts; and D1, D2... Dn = number of days elapsed from sowing to the first, second and last counts.

The data obtained was subjected to normality and homogeneity tests and analysis of variance using the SISVAR computer programme (FERREIRA, 2011). The means were compared using the F test and Tukey's test, respectively, both at the 5% probability level (GOMES, 2000). In order to obtain greater reliability, the experiment was repeated three times and, at the end, a joint analysis of the data was carried out.

RESULTS AND DISCUSSION

There was a significant effect only for the osmoconditioning solution factor and the data can be seen in Table 1. From the results obtained in the germination test, it can be seen that the osmoconditioning treatments using KNO3 solution provided the highest germination percentages up to 70 days. These results prove the efficiency of osmoconditioning araçá-boi seeds in KNO3 solution compared to the other treatments, with 99% of seeds germinating.

Table 1 - Average germination percentage and germination speed index (GVI) values of araçá-boi seeds 70 days after osmoconditioning in C12H22O11, KNO3, C12H22O11 + KNO3 and H2O, Boa Vista, RR, 2015

Table 1 - Average values of germination percentage and germination speed index (GSI) of Araçá-boi seeds 70 days after submitted to osmopriming in $C_{12}H_{22}O_{11}$, KNO_3, $C_{12}H_{22}O_{11}$ + KNO3 and H_2O, Boa Vista, RR, 2015

Solution	Germination percentage
KNO3	99 a
C12H22O11	37 b
C12H22O11 + KNO3	27 c
H2O	7 d
	Germination speed (Index)
KNO3	2,14 a
C12H22O11	1,53 b
C12H22O11 + KNO3	1,51 b
H2O	0,04 c

Averages followed by the same lowercase letter in the column and uppercase letter in the row do not differ according to Tukey's test at the 1% probability level.
The means followed by the same lower case letter in the column and upper case on the line, do not differ by Tukey test at 1% probability.

The behaviour of the germination rate (GVR) of the araçá-boi seeds during the 70-day evaluation period, conditioned in the different treatments with osmotically active solutions and water, can be seen in Figure 1. The IVG was higher in seeds osmoconditioned in KNO_3 solution. These results are similar to the behaviour observed for germination percentage.

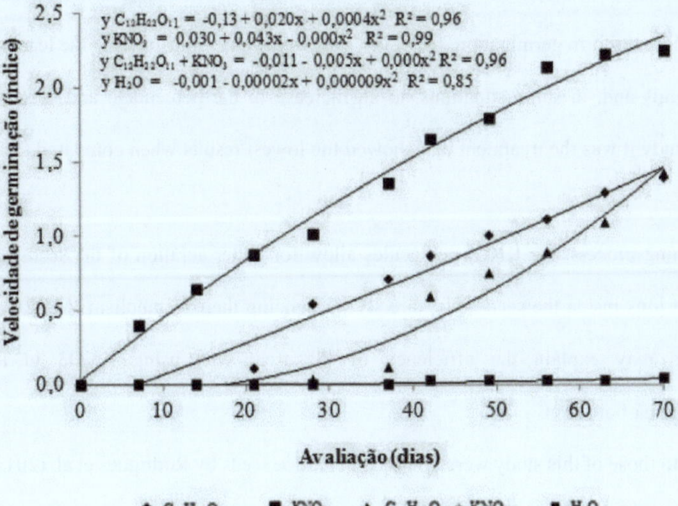

Figure 1 - Cumulative germination rate (index) of araçá-boi seeds osmoconditioned in $C_{12}H_{22}O_{11}$,

KNO_3, KNO_3 + $C_{12}H_{22}O_{11}$ and H_2O up to 70 days.

Figure 1 - Cumulative germination speed (index) of Araçá-boi seeds osmoprimed in $C_{12}H_{22}O_{11}$, KNO_3, KNO_3+ $C_{12}H_{22}O_{11}$ and H_2O up to 70 days.

The main characteristic of the araçá-boi seeds used in this study is their long germination period, which can take up to 8 months to reach 100% germination. In this study, it was found that osmoconditioning in a KNO_3 solution was effective in promoting the germination of araçá-boi seeds, showing a higher germination rate and was significantly superior to the other treatments.

It was also found that the seeds treated with KNO_3 osmoconditioning solution started germinating 7 days after the experiments were set up, followed by the sucrose treatment. On the other hand, seeds subjected to conditioning in water alone began to germinate between 42 and 49 days after installation (Figure 1), reaching a maximum of 7% germination (Table 1). As a result, not only did osmoconditioning with KNO_3 result in a higher germination percentage, but it also favoured a higher germination speed.

Sert et al. (2009) found that water leaching can be an efficient method for removing dormancy caused by inhibitors. Castro and Hilhorst (2004) also commented that soaking seeds in water promotes faster and more uniform germination. However, although water contributes to the leaching of inhibitory compounds and, in some situations, to an increase in the percentage and speed of germination, in this study it was the treatment that showed the lowest results when compared to the other treatments.

In the conditioning process, the KNO_3 molecules allow for better aeration of the seeds and penetration of the salt ions inside the seed cells, directly influencing their metabolism (MARCOS FILHO, 2005). This may explain the efficiency of the treatments using KNO_3 in the osmoconditioning of araçá-boi seeds.

Results similar to those of this study were obtained in lettuce seeds by Rodrigues et al. (2012). These authors observed that osmoconditioned seeds showed an increase in germination rate. In experiments carried out with pepper seeds, osmoconditioning in KNO_3 also favoured germination speed, especially in seeds harvested from fruit at the beginning of the ripening phase (ALBUQUERQUE et al., 2009). Contrary results have also been observed in the literature in beetroot (COSTA and VILELA, 2006) and onion seeds (CASEIRO and MARCOS FILHO, 2015), whose performance and seed germination were reduced after osmoconditioning. This shows that the efficiency of the osmoconditioning process is possibly species-dependent.

An increase in germination efficiency promoted by osmoconditioning in low vigour melon and carrot seeds was observed by Nascimento and Aragão (2004) and Pereira et al. (2008) respectively. On the other hand, lettuce and coffee seeds with high viability, when subjected to osmoconditioning, did not show an increase in germination efficiency (FESSEL et al., 2001; PERTEL et al., 2001).

Depending on the species, germination speed may not be influenced by pre-soaking, which is a treatment that does not necessarily accelerate the germination speed of seeds of all species (NASCIMENTO, 2004). This result could be related to the characteristics of each species, batch and other morphogenetic properties of the seeds, although osmotic conditioning has more significant effects on seeds of low and medium vigour, such as the seeds used in this study. These seeds are considered to be of low vigour, as the seeds immersed and soaked in distilled water did not germinate more than 7% (Table 1 and Figure 1) in the 70 days of monitoring.

CONCLUSIONS

The osmoconditioning of araçá-boi seeds in a KNO3 solution is effective in promoting germination and increasing germination speed.

Araçá-boi seeds osmoconditioned in KNO3 show a germination percentage of 99% up to 70 days.

The period of osmoconditioning did not influence the germination of araçá-boi seeds.

CHAPTER 3

BIOSPECKLE AS AN AUXILIARY TOOL FOR ASSESSING THE DORMANCY OF ARAÇÁ- BOI SEEDS

SUMMARY

The araçá-boi is a fruit native to the Amazon known for the organoleptic qualities of its fruit, but the major problem presented by this fruit is the seed's intrinsic dormancy. The use of coherent light in the assessment of biological activity has great potential as it makes it possible to identify damage to seeds. This technique is based on the phenomenon known as dynamic speckle or biospeckle. The aim was to study the use of the biospeckle technique to determine the viability and vigour of seeds and correlate this with germination behaviour and seedling development in the field. In experiment 1, the seeds used were separated into large, medium and small and subjected to treatments to break dormancy: 1- control, 2- removal of the tegument using sandpaper and 3- removal of the tegument using a scalpel. In experiment 2, uniform seeds were used and to break dormancy the seeds were placed in gerboxes previously lined with filter paper and Potassium Nitrate solution for osmoconditioning for 24 hours. After the treatments used in the two experiments, images were taken in order to use the biospeckle technique. The use of the biospeckle technique was efficient in determining the viability and vigour of araçá-boi seeds. Large and medium-sized seeds showed higher activities. The emergence of large and medium seeds, the speed of emergence and the development of seedlings were related to the activity observed by the biospeckle technique. Seeds osmoconditioned in KNO3 for 24 hours showed higher activities. The emergence of seeds osmoconditioned in KNO3 for 24 hours and the speed of emergence were related to the activity observed by the biospeckle technique.

Keywords: Eugenia stiptata; Germination; Laser; Dynamic Speckle;

INTRODUCTION

The araçá-boi is a fruit species native to the Amazon with great potential for agro-industrial utilisation, as it has fruit with good physical and chemical characteristics and sensory attributes that are acceptable (ROGEZ et al., 2004).

The great difficulty in growing this fruit tree is the long time it takes for the seeds to germinate and the seedlings to emerge, which is related to the inherent dormancy of the seeds. This dormancy may be related to the hardness and impermeability of the tegument to water and gases, the immature embryo and the presence of germination inhibitors (GENTIL and FERREIRA, 1999; SERT et al.,

2009).

Another important factor is the size of the seeds, since this characteristic influences the percentage of germination and emergence, as well as their germination speed (COSTA et al., 2006; OLIVEIRA et al., 2003).

The search for methods to assess seed viability in a non-destructive and effective way has led to the study of promising new techniques. Biospeckle or dynamic speckle is a phenomenon known as the use of coherent light to assess the biological activity of living tissues or to study the dynamics of particulate systems.

This phenomenon can be noticed when a beam of coherent light is scattered when it hits a surface that exhibits some kind of dynamic activity (RODRIGUES, 2007).

A biospeckle is an interference pattern formed by the diffuse reflection of coherent light scattered when it interacts with an object that shows some kind of activity, biological or otherwise. The interference pattern changes over time due to the structures responsible for the scattering being active (OLIVEIRA et al., 2009).

According to Rodrigues et al. (2007), if the diffusing object changes, so do the individual grains of the speckle pattern. This suggests that the biospeckle patterns contain information about the movement of the object.

To analyse the level of activity of biological material, what has been proposed with the use of lasers is to analyse the information taken from the autocorrelation function of the Spatial Temporal Speckle (STS) as presented by Rabelo (2000) and Romero (1999). This tool has also been used by Bergkvist (1997), Oulamara (1989) and Xuet al. (1995).

Another way of analysing STS properties is to calculate the intensity dispersion modulus, also known by some authors as the moment of inertia (mi), which is obtained using the Occurrence Matrix (MOC) proposed by Arizaga et al. (1999).

Due to the complexity of biospeckle, it would be very convenient to have a model that describes at least the main characteristics of the phenomenon (BRAGA JÚNIOR, 2000). The pattern generated by the biospeckle is dynamic and random, and must be analysed using pattern processing techniques and statistical treatment, since visual observation only allows the existence of the phenomenon to be identified, but does not allow it to be quantified (RABAL et al., 1998).

Biospeckle is a valuable tool for studies in various areas of knowledge, especially agronomy, and can be used to assess the germination potential of seeds (Silva, 2007).

Braga Júnior et al., (2001), analysed seeds using the biospeckle technique and stated that the use of lasers has great potential, as it makes it possible to identify damage to seeds in order to assess viability.

In studies carried out on bean seeds (Phaseolus vulgaris L), Rodrigues et al. (2007) separated live and dead tissues using the biospeckle technique. Botega et al. (2009) emphasised that the incidence of the laser did not significantly affect leaf activity, confirming experiments by Seizt (1979) and Rabelo (2000) who assessed the influence of the laser on the metabolism of oranges, showing that tissue subjected to a low-power laser did not alter the biospeckle results over time.

The aim of this work was to study the use of the biospeckle technique to determine the viability and vigour of seeds and correlate this with the germination behaviour and development of seedlings in the field. To this end, two experiments were carried out to assess the possibility of using the biospeckle technique as an auxiliary tool to reduce the time taken to assess the treatment for overcoming dormancy in araçá boi seeds.

MATERIAL AND METHODS

Material collection

The seeds used in this study came from the fruits of various populations of E. stipitata, collected in the city of Manaus between 7 and 9 November 2013. The seeds were extracted manually, followed by washing in running water using a sieve to completely remove the mucilage. The seeds were then left to dry for 2 hours at an average temperature of 25°C.

EXPERIMENT 1

Choosing seeds

The seeds were quantified and separated according to their size into three groups: 1 - Large, seeds with a mass greater than 0.78g, 2 - Medium, seeds with a mass between 0.77g and 0.19g and 3 - Small, seeds with a mass less than 0.18g.

Treatments to break dormancy

Ninety-six seeds were selected from each group (Small, Medium and Large) and numbered sequentially from 01 to 288. The seeds were subjected to the following treatments to break dormancy: intact seed (Test), removal of the tegument using sandpaper and removal of the tegument using a scalpel, after which they were submerged in water for 12 hours. After this period, the seeds were taken to the Plasma and Atomic Spectroscopy Laboratory (LaPEA) at the Federal University of Roraima to acquire the images in order to use the biospeckle technique.

EXPERIMENT 2

Choosing seeds

To carry out this experiment, the seeds were all selected with the same size, as uniformly as possible.

Treatments to break dormancy

80 seeds were selected uniformly. Forty seeds were separated from the samples used as controls and then immersed in a 2% NaClO solution for 60 minutes to disinfect them. Afterwards, they were washed in distilled water and placed in gerboxes previously lined with filter paper and Potassium Nitrate solution (KNO3 - osmotic potential of -1.1MPa and its concentration was based on the Van't Hoff equation (Hillel, 1971)) for osmoconditioning for 24 hours. Osmoconditioning was carried out in B.O.D (Biochemical Oxygen Demand) at 10°C. The gerboxes were wrapped in plastic bags to prevent the solution from evaporating. After treatment, the seeds were washed in distilled water and then submerged in water for 12 hours. After this period, the seeds were taken to the Plasma and Atomic Spectroscopy Laboratory (LaPEA) at the Federal University of Roraima to acquire the images in order to use the biospeckle technique.

Image acquisition in the experiments

The seeds were removed from the water, and a period of 5 minutes was allowed for the loss of excess moisture, after which the laser reading was carried out. Image acquisition time was around 12 seconds, with each treatment comprising 32 seeds, and at the end of all the readings 288 videos were obtained.

The experimental apparatus used consisted of a He-Ne laser with a wavelength of 632 nm, a CCD camera, a biconvex lens for diverging the laser to ensure that the entire seed was illuminated, and a microcomputer. Figure 1 shows a schematic of the experimental apparatus used.

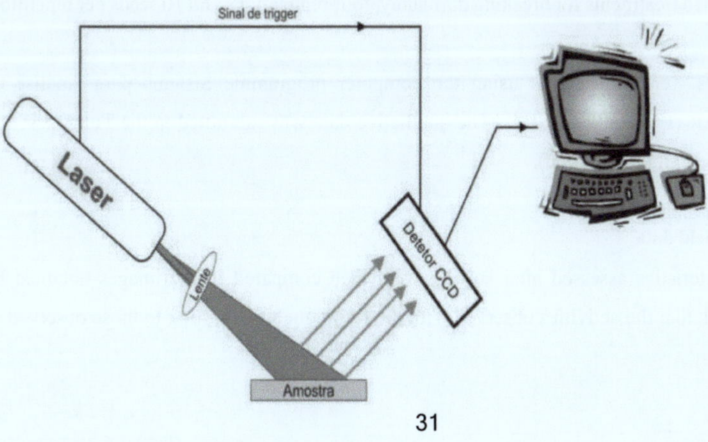

Figure 1: Schematic of the experimental apparatus used at LaPEA.

Interpreting images

Each 12-second video was converted into 94 images. Once these images were obtained, they were edited using the Gimp programme.

The images were compared two by two, superimposed, so that in the end only the difference between them was obtained. After editing these images, 15 patterns of differences were obtained for each seed and, using these differences, it was possible to determine the changes that had occurred in three randomly chosen bands of pixels in the images.

To interpret the results, it was established that the region that showed a difference in activity in the image was given a value of 1 (white colour) and the region that showed no difference, i.e. no activity, was given a value of 0 (black colour).

Sowing

After acquiring the images in experiment 1, the seeds were sown in a seedbed containing sand + sawdust (1:1), and assessed for the percentage of emergence and the speed of emergence (index), the length of the aerial part using a millimetre ruler in cm and the number of leaves. At the end of 100 days, the length of the root system, the dry mass of the aerial part and the root system were assessed using a digital scale with a precision of 0.001g.

The experimental design used was entirely randomised, in a 3 x 3 factorial scheme, with three seed sizes and three dormancy-breaking treatments, with 9 treatments, four repetitions and 8 seeds per repetition, totalling 288 seeds.

In experiment 2, the seeds were sown in a seedbed containing a sand + sawdust (1:1) substrate, and assessed for percentage emergence and speed of emergence (index). The design used was entirely randomised, with two treatments for breaking dormancy, four replications and 10 seeds per repetition, totalling 80 seeds.

The analyses were carried out using the computer programme Sistema para Análise de Variância - SISVAR (FERREIRA, 2011). The qualitative data was submitted to the Tukey test at a 1% probability level (GOMES, 2000).

Comparison with field data

The characteristics assessed after sowing were then compared to the images obtained by biospeckle to check that the activities observed with the technique were similar to those observed in the field.

RESULTS AND DISCUSSION

Figure 2 shows the difference in activity between large and small araçá-boi seeds.

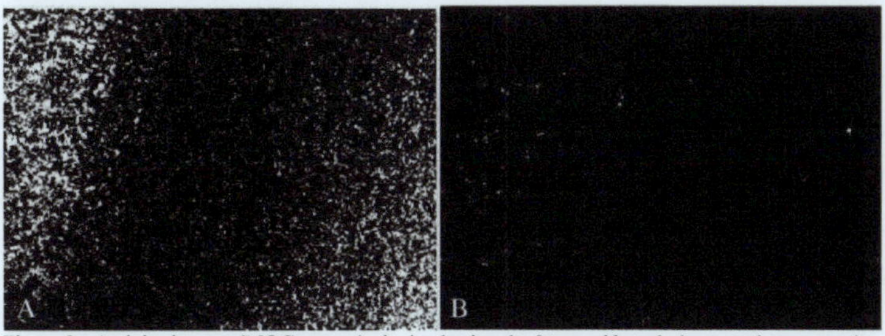

Figure 2 - Activity images (MOC pattern) obtained using the *biospeckle* technique used to assess the viability of large (A) and small (A) araçá-boi seeds.

The highest sequence of activity observed in the images generated by the biospeckle technique in experiment 1 was obtained by the large and medium-sized araçá-boi seeds, while the small seeds showed low activity. The treatments used to accelerate the seed germination process had no influence on the activities shown (Figures 3).

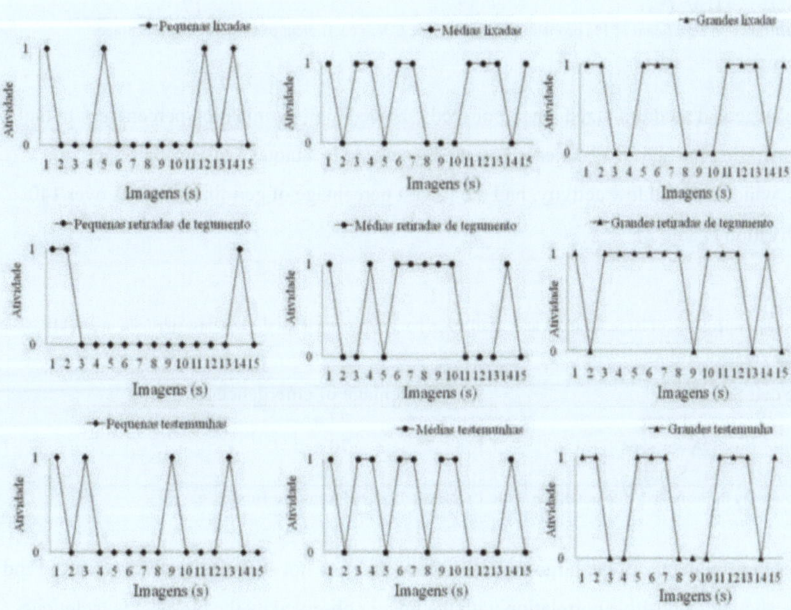

Figure 3 - Activities shown by small, medium and large seeds after treatments to accelerate the germination process, recorded using the biospeckle technique, Boa Vista, RR.

The images obtained by this technique indicate the biological activity index of the seeds and can be used to assess their germination potential (SILVA, 2007).

According to Braga Júnior et al. (2001), the use of lasers has great potential for identifying seed damage in order to assess viability. This information corroborates studies carried out on bean seeds (Phaseolus vulgaris L) by Rodrigues et al. (2007). The authors were able to separate living and dead tissues using the biospeckle technique.

It can be seen that the small seeds had low activity and consequently low germination rates. This result is related to the behaviour of the seeds in the field, where it was observed that the seed sizes were significant in the germination percentage, and the treatments used to break dormancy also had no influence on germination in the field (Table 1).

Table 1 - Mean square values and their respective significances for the emergence data of large, medium and small araçá-boi seeds after dormancy-breaking treatments, Boa Vista, RR, 2014

Variation Factors	Degrees of Freedom	Mean Squares
Treatments	2	506.77 ^{ns}
Sizes	2	19272,44**
Treatments x Sizes	4	89,65$^{n.s}$
Waste	27	172,34
Total	35	
C.V. (%)		17

$^{n.s}$, **, not significant and significant at 1% probability by the F test, C.V.= coefficient of variation in percentage.

The large and medium-sized araçá-boi seeds showed high germination percentages in the field, demonstrating the activities observed by the biospeckle technique. On the other hand, the small seeds, which showed low activity, had the lowest percentage of germinated seeds over 140 days of cultivation (Table 2).

Table 2. Percentage emergence of large, medium and small araçá-boi seeds sown in sand + sawdust substrate over 140 days of cultivation, Boa Vista, RR, 2014

Seed Sizes	Percentage of emergence
Large	77 a
Averages	70 a
Small	3b

Averages followed by the same letter do not differ at the 1% probability level using the Tukey test.

The emergence percentages of the large and medium seeds were not statistically different at the end of 140 days, showing a positive correlation with the images observed by the biospeckle technique, which showed no differences between the activities of the large and medium seeds. The large and medium seeds also had a higher speed of seedling emergence, showing good development in the field (Figure 4).

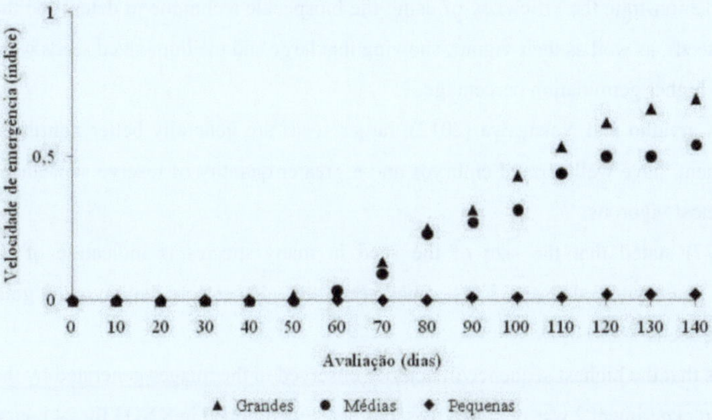

Figure 4 - Cumulative emergence rate (index) of seedlings obtained from large, medium and small araçá-boi seeds up to 140 days.

The large and medium seeds showed similar emergence speeds until approximately 60 days of cultivation, and from this date onwards the emergence speed of the large seeds was higher. This result did not influence the emergence percentages of the large and medium seeds, which did not differ statistically at the end of 140 days of cultivation (Table 2).

In studies on the influence of seed size on the early germination of bacuripari (Rheedia gardneriana), Oliveira et al. (2003) observed higher germination speeds and percentages for larger seeds. Higher emergence rates were also found for large and medium-sized red jamb (Sysygium malaccense) seeds (COSTA et al., 2006).

The characteristics assessed up to 140 days of cultivation showed that large and medium seeds not only had higher germination percentages, but also better seedling development (Table 3).

Table 3. Aerial part length (APL), root system length (RSL), aerial part dry mass (APDM), root system dry mass (RSDM) and number of leaves (NF) of araçá-boi leaves emerged up to 140 days of cultivation in sand + sawdust substrate, after viability analysis by biospeckle, Boa Vista, RR, 2014

Size of Seeds	CPA (cm)	CSR (cm)	MSPA (g)	MSSR (g)	NF
Large	15,61 a	12,85 a	0,210 a	0,110 a	20 a
Averages	11,91 b	11,17 a	0,120 b	0,060 b	18 a
Small	0,92 c	1,25 b	0,003 c	0,002 c	3b
C.V (%)	3,89	4,01	3,32	2,09	2,28

Averages followed by the same letter do not differ by Tukey's test at 1% probability.

The seedlings that emerged from large seeds had longer aerial part lengths and higher aerial part and root dry masses. However, in terms of the length of the root system and the number of leaves, the seedlings from large and medium seeds did not differ statistically, thus showing uniformity.

These results demonstrate the efficiency of using the biospeckle technique to determine the viability of araçá-boi seeds, as well as their vigour, showing that large and medium-sized seeds were more viable and had a higher germination percentage.

According to Carvalho and Nakagawa (2012), larger seeds are generally better nourished during their development, have well-formed embryos and a greater quantity of reserve substances and are therefore the most vigorous.

Popinigis (1977) stated that the size of the seed in many species is indicative of its physiological quality. Thus, within the same batch, small seeds show lower germination and vigour than medium and large seeds.

Figure 5 shows that the highest sequence of activity observed in the images generated by the biospeckle technique in experiment 2 was obtained by seeds osmoconditioned in KNO3 for 24 hours, while the control seeds showed low activity.

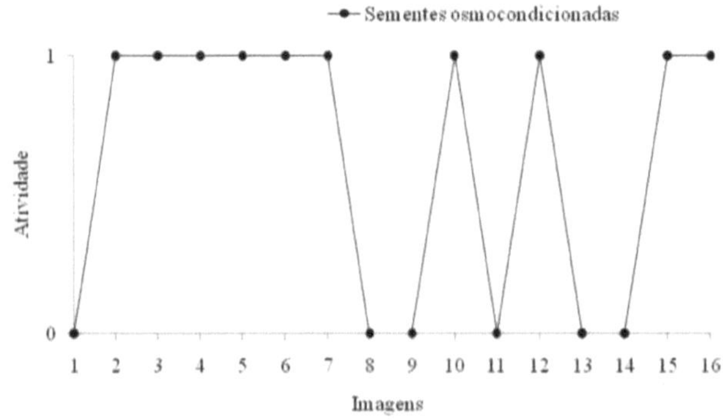

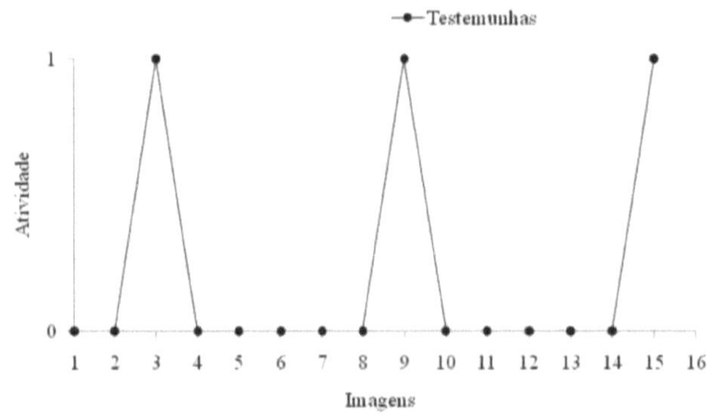

Figure 5 - Activities shown by seeds osmoconditioned in KNO3 for 24 hours and seeds used as controls, recorded using the biospeckle technique, Boa Vista, RR.

This result is related to the behaviour of the seeds in the field, where it was observed that osmoconditioned seeds had a higher germination percentage. According to Silva (2007), the images obtained by the biospeckle technique indicate the biological activity index of the seeds and can be used to assess their germination potential.Table 4 shows that there was a significant effect for the treatment used to break seed dormancy.

Table 4: Mean square values and their respective significances for the emergence data of seeds osmoconditioned in KNO3 for 24 hours, Boa Vista, RR, 2014

Sources of Variation	Degrees of Freedom	Mean Squares
Treatments	1	19012,5**
Waste	6	12,5$^{n.s}$
Total	7	
C.V (%)	7	

$^{n.s}$, **, non-significant and significant at 1% probability by the F test, C.V.= coefficient of variation in per cent.

It can be seen that the seeds osmoconditioned in KNO3 for 24 hours showed high activity and consequently high germination rates. There was a statistical difference between the araçá-boi seeds osmoconditioned in KNO3 for 24 hours and the control (Table 5).

Table 5. Germination percentage of araçá-boi seeds osmoconditioned in KNO_3 for 24 hours sown in sand + sawdust substrate over 63 days of cultivation, Boa Vista, RR, 2014

Treatments	Emergency Percentage
Osmocondicioned	100 a
Witnesses	2,5 b

Averages followed by different letters differ statistically by the Tukey test at 1% probability.

The emergence percentage of the osmoconditioned seeds differed statistically from the control seeds, reaching 100% emergence after 63 days, showing a positive relationship with the images observed by the biospeckle technique, which showed differences between the activities presented by the osmoconditioned seeds and the control seeds.

In the osmoconditioning process, as well as allowing the seeds to be better aerated, the KNO3 molecules allow the salt ions to penetrate inside the seed cells, directly influencing their metabolism (MARCOS FILHO, 2005).The speed of emergence of the osmoconditioned seeds was proportional to their percentages of emergence (Table 5).

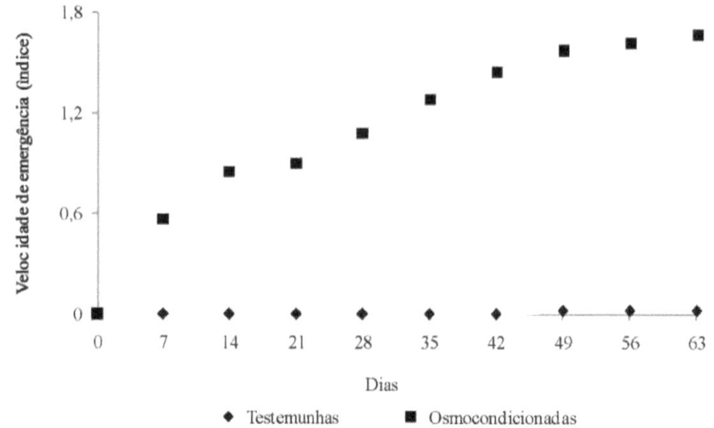

Figure 6 - Emergence rate of araçá-boi seeds osmoconditioned in KNO_3 for 24 hours, obtained after 63 days.

The osmoconditioned seeds also had a faster seedling emergence rate, demonstrating the positive relationship between the images obtained by the Biospeckle technique and the behaviour of the seeds in the field.

Osmoconditioning consists of the controlled hydration of seeds in an osmotic solution, stimulating their metabolism without, however, allowing the primary root to emerge (NASCIMENTO, 2004). This conditioning activates the degradation, translocation and assimilation of reserves, allowing the seeds to reach a relatively uniform metabolic state when the water supply is interrupted.

CONCLUSIONS

The use of the biospeckle technique is efficient in determining the viability and vigour of araçá-boi seeds. Large and medium-sized seeds show higher activity.

The emergence of large and medium seeds, the speed of emergence and the development of seedlings are related to the activity observed by the biospeckle technique.

Seeds osmoconditioned in KNO3 for 24 hours showed higher activities.

The emergence of seeds osmoconditioned in KNO3 for 24 hours and the speed of emergence are related to the activity observed by the biospeckle technique.

BIBLIOGRAPHICAL REFERENCES

ALBUQUERQUE, K. S.; GUIMARÃES, R. M.; GOMES, L. A. A.; VIEIRA, A. R.; JÁCOME, M. F. Osmotic conditioning of chilli seeds. **Revista Brasileira de Sementes**, v. 31, n. 4, p.100-109, 2009.

ANDRADE, R. A.; JESUS, N.; MARTINS, A. B. G. Soaking and germination of Camu-camu seeds.

Acta Scientiarum Agronomy, p.499-501, 2006.

ANJOS, A. M. G.; FERRAZ, I. D. K. Morphology, germination and water content of araçá-boi seeds (Eugenia stipitatassp. Sororia). **Acta Amazônica**, v.29, n.3, p.337-348, 1999.

ARAUJO, P. C.; TORRES, S. B.; BENEDITO, C. P.; PAIVA, E. P. Physiological conditioning and vigour of maxixe seeds. **Revista Brasileira de Sementes**, v. 33, n. 3, p. 482-489, 2011.

ASAKURA, T., TAKAI, N. Dynamic laser speckles and their application to velocity measurements of the diffuse object. Journal of Applied Physics, v. 25, p. 179-194, 1981.

BASKIN, C.C.; BASKIN, J.M. Seed: ecology, biogeography and evolution of dormancy and germination. San Diego: **Academic Press**, 1998, 666p.

BERGKVIST, A. Biospeckle-based study of the line profile of light scattered in strawberries. 62 f. Dissertation (Master in Science) - Faculty of Technology at Lund University, La Plata, 1997.

BEWLEY, D. D.; BLACK, A. M. **Seeds: physiology of development and germination**. NewYork: Plenum, p 445, 447, 1994.

BORGES, R. C. F.; COLLAÇO JUNIOR, J. C.; SCARPARO, B.; NEVES, M. B.; CONEGLIAN, A. Characterisation of the soaking curve of jatropha seeds. **Revista científica eletrônica de engenharia florestal**, v.8, n.13, p.1-8, 2009.

BOTEGA, J. V. L, BRAGA JR, R. A.; MACHADO, M. P. P.; LIMA, L. A.; FARIA, R. O. Evaluation of laser interference in living biological material. **Brazilian Journal of Agricultural and Environmental Engineering**, v.13, n.4, p.483-486, 2009.

BRAGA JR, R. A.; BORÉM, F. M.; RABAL, H. J.; TRIVI, M. R.; ARIZAGA, R.; DAL FABBRO, I. M.; SALEH, B. B. Evaluation of the influence of the humidity of bean seeds (Phaseolusvulgaris L) on the analysis by the laser biospeckle technique. **Engineering Agrícola**, Jaboticabal, v.21, n.1, p.101-106, jan. 2001.

BRAGA JR., R.A... Seeds characterisation by dynamic speckle patterns: a proposal. In: Environmental and Industrial Sensing Conference, Boston. **Proceedings**...Boston: SPIE, 2000, 1 CD Rom.

BRAZIL. Ministry of Agriculture and Agrarian Reform. Rules for analysing seed. National Secretariat for Agricultural Defence. National Department of Plant Production. Plant Laboratory

Coordination. Brasília, DF, 365p, 2009.

CARDOSO, G.L;LOMÔNACO, C.Phenotypic variations and plastic potential of Eugenia calycina Cambes . (Myrtace) in a cerrado-vereda transition area. **Revista Brasileira de Botânica**, v.26, n.1, p.131-140, 2003.

CARDOSO, V. J. M. Germination. In: Kerbauy, G. B., **Fisiologia Vegetal**. Guanabara- Koogan, C. 17, p.386-407, 2004.

CARVALHO, N. M.; NAKAGAWA, J. Seeds: science, technology and production. 4.ed. Jaboticabal: FUNEP, p.588, 2000.

CARVALHO, N. M.; NAKAGAWA, J. Seeds: science, technology and production. Jaboticabal, FUNEP, 590p., 2012.

CASEIRO, R.F.; MARCOS FILHO, J. Methods for drying onion seeds submitted to physiological conditioning. **Horticultura Brasileira**, v.23, n.4, p.887- 892, 2005.

CASTRO, R. D.; HILHORST, H. W. M. Soaking and Reactivation of Metabolism. In: Ferreira, A. G. and Borghetti, F. **Germination: from basic to applied**. Artimed, Porto Alegre, p.149-162, 2004.

CHAVES FLORES, W. B.; CLEMENTE, C. R. Considerations on the araçá-boi (Eugenia Stipitata McVaugh, Myrtaceae) in the Brazilian Amazon. In: **BRAZILIAN FRUIT CONGRESS**. Florianópolis. Proceedings... Florianópolis: SBF, 1984. p.167-177.

CLEMENT, C. R. et al. On-farm conservation. In: NASS, L. L. (Org.). plant genetic resources. Brasília: Embrapa Recursos Genéticos e Biotecnologia, 2007.

CLEMENT, C.R. ARKCOLL, D.B.. 1989. The pejibaye palm: economic potential and research priorities. The pejibaye palm: economic potential and research priorities. In: Wickens, G., N. Haq & P. Day (eds). New Crops for Food and Industry. Chapman & Hall, New York. p. 304-322.

COSTA, C. J.; VILLELA, F.A. Osmotic conditioning of beetroot seeds. **Revista Brasileira de Sementes**, v.28, n.1, p.21-29, 2006.

COSTA, N.P.; BRUNO, R. L. A.; SOUZA, F. X.; LIMA, E. D. P. A. Fruit ripeness stage and endocarp pre-soaking time in the germination of umezeiro (Spondias tuberosa Arr. Câm.) seeds. **Revista Brasileira de Fruticultura**. Jaboticabal, v.23, n.3, p.738- 741, 2001.

COSTA, R. S.; OLIVEIRA, I. V. M.; MÔRO, F. V.; MARTINS, A. B. G. Aspectos morfológicos e influência do tamanho da semente na germinação do jambo-vermelho. **Revista Brasileira de Fruticultura**, Jaboticabal, v. 28, n. 1, p. 117-120, 2006.

FERREIRA, D. F. Sisvar: a computer statistical analysis system. **Ciência e Agrotecnologia,** Lavras, v.35, n.6, p.1039-1042, 2011.

FERREIRA, G.; GUIMARÃES, V. F; PINHO, S. Z.; OLIVEIRA, M. C.; RICHART A, BRAGA, J. F.; DIAS, G. B. Water absorption curve in seeds of atemoya (Annona cherimola Mill. X Annona squamosa L.) cv. gefner. **Revista Brasileira Fruticultura,** v.28, n.1, p.121-124. 2006.

FERREIRA, W.R.; RANAL, M.A. Seed germination and seedling growth of Brassica chinensis L. var. Parachinensis (Bailey) Sinskaja (Malaysian cabbage). **Pesquisa Agropecuária Brasileira,** v.34, n.3, p.353-361. 1999.

FESSEL, S.A.; VIEIRA, R.D.; RODRIGUES, T.J.D.; FAGIOLI, M.; PAULA, R.C. Efficiency of osmotic conditioning in lettuce seeds. **Revista Brasileira de Sementes,** v.23, n.1, p. 128-133, 2001.

FREITAS, R. R.; CARVALHO, D. A.; ALVARENGA, A. A. Breaking the dormancy of marmalade grass seeds. **Revista Brasileira de Fisiologia Vegetal,** v.2, n.2, p. 31-35, 1990.

GENTIL, D. F. O.; FERREIRA, S. A. N. Viability and overcoming dormancy in araçá-boi seeds (Eugenia stipitatassp. Sororia). **Acta Amazônica,** v.29, n.1, p.21-31, 1999.

GOGOSZ, A.M.; COSMO, N.L.; BONA, C.; SOUZA, L.A. Morphoanatomy of CampomanesiaxanthocarpaO. Berg. (Myrtaceae). **Acta Botanica Brasilica**, v.24, n.3, p.613-623, 2010.

GOMES, F. P. Curso de estatística experimental 14 ed. Piracicaba, 477 p, 2000.

HAIG, D. WESTOBY, M. Seed size, pollination casts and angiosperm success. Evolutionary Ecology, London, v. 5, p. 231-247, 1991.

HILLEL, D. Soil and water: physical principles and processes. New York: Academic Press, p.288, 1971.

HISIÃO, A. L.; QUICK, W. A. Actions of sodium hypochlorite and peroxide on seed dormancy and germination of Will oats, Avenafátua L. **Weed Research**, Oxford, v.24, n.6, p.411-419, 1984.

KISSMANN, C.; SCALON, S.P.Q.; SCALON FILHO, H.; VIEIRA, M.C. Bioregulator and osmotic preconditioning in seed germination and initial growth of carobinha seedling *(JacarandadecurrenssUbsp. symmetrifaliolataFanas* & Proença) - Bignoniaceae. **Revista Brasileira de Plantas Medicinais**, v.13, n.1, p.58-67, 2011.

LIMA, J. D; ALMEIDA, C. C; DANTAS, V. A. V; SILVA, B. M.DA; MORAES, W. S. Effect of temperature and substrate on seed germination of *caesalpinia ferrea* mart. extul. (leguminosae, caesalpinoideae). **Revista Árvore,** v.30, n.4, p.513-518, 2006.

MAGUIRE, J. D. Speed of germination-and in selection and evaluation for seeding emergence and vigour. **Crop Science**, v. 2, n. 2, p. 176-177, 1962.

MALTA, M. R.; PEREIRA, R. G. F. A.; CHAGAS, S. J. R. Electrical conductivity and potassium leaching from coffee bean exudate: some factors that may influence these evaluations. **Ciência e Agrotecnologia**, Lavras, v.29, n.5, p.1015- 1020, 2005.

MARCOS FILHO, J. Fisiologia de sementes de plantas cultivadas. Piracicaba: FEALQ, 495p, 2005.

MARQUES, M. A.; PAULA, R. C. RODRIGUES, T. J. D. Suitability of the electrical conductivity test to determine the physiological quality of jacaranda da-bahia seeds (Dalbergia nigra (Vell.) Fr. All. Ex. Benth). **Revista Brasileira de Sementes**, Londrina, v.24, n.1, p.271-278, 2002.

MCVAUGH, R. Tropical American Myrtaceae. Notes on generic concepts and descriptions of previously unrecognised species. **Fieldiana, Botany**, v.29, n.3, p.145- 228, 1956.

MENDES, A. M. S. Eugenia stipitata ssp. Sororia McVaugh (araçá-boi) - Myrtaceae: physiological and morphoanatomical approach to seed, germination and seedling. PhD Thesis, Federal University of Amazonas, 117p, 2011.

MENDONÇA, M. S.; BARBOSA, T. C. T. S.; ARAÚJO, M. G. P.; OLIVEIRA, M. G. P. Floral morphology of some fruit trees found in Manaus. Manaus: EDUA, 2001. 56.p.

NASCIMENTO, W. M. Osmotic conditioning of vegetable seeds. Brasília: **Embrapa**, **Technical Circular**, v. 33, 12p., 2004.

NASCIMENTO, W. M.; ARAGÃO, F. A. S. Muskmelonseedpriming in relationtoseed vigour. **Scientia Agricola**, v. 61, n. 1, p. 114-117, 2004.

NASCIMENTO, W. M.; PEREIRA, R. S. Prevetingtermohinibition in carrot by seed priming. **Seed Science Technology**, v. 35, n. 1, p. 503-506, 2007.

OLIVEIRA, I.V.M.; ANDRADE, R.A.; MARTINS, A.B.G. Influence of seed size on early germination of bacuripari (Rheediagardneriana). In: Annual Meeting of the Inter-American Society of Tropical Horticulture, Fortaleza. Proceedings... p.88. 2003.

OLIVEIRA, J. R.; DUARTE, N. F.; GONÇALVES, L. D.; SOUZA, F. V. P.; SILVA, J. C. O.; SILVA, U. T. G. Aplicabilidade do laser Biospeckle nas Ciências Agrárias.In: II Semana de Ciência e Tecnologia IFMG, Campus Bambuí, 2009.

OULAMARA, A.; TRIBILLON G.; DUVERNOY J.; ARIZAGA, R.; TRIVI, M.R.; RABAL, H. J. Speckle time evolution characterisation by co-occurrence matrix analysis. **Optics and Laser Technology,** v. 4-5, n.3, p.1-7, 1999.

PEREIRA, M.D.; DIAS, D.C.F.S.; DIAS, L.A.S.; ARAÚJO, E.F. Germination and vigour of carrot seeds osmoconditioned in moistened paper and aerated solution. **Revista Brasileira de Sementes**, v.30, n.2, p.137-145, 2008.

PERTEL, J.; DIAS, D.C.F.S.; DIAS, L.A.S.; ALVARENGA, E.M. Effect of physiological conditioning on the germination and vigour of coffee seeds *(Coffeaarabica* L.). **Revista Brasileira de Armazenamento**, Especial, n.3, p.39-45, 2001.

PICÓN, B. C. Araza cultivation. Iquitos, INIA, p.8, 1989.

PINEDO, P. M.; RAMIREZ, N.; BLASCO, M. L. Notas preliminares sobre el araza (Eugenia stipitata), frutal nativo de la Amazonia Peruana, Pub. Misc. 229, **Instituto Nacional de Investigación Agrária**, Lima, Peru. 1981, 58p.

PINTO, W.S;DANTAS, A. C. V.; FONSECA, A. A. O; LEDO, C. A. S. ; JESUS, S. C. CALAFANGE, P. L. P.; ANDRADE, E. M. Physical, physico-chemical and chemical characterisation of fruit from cajazeiras genotypes. Pesquisa Agropecuária Brasileira, Brasília, v. 38, p. 1059-1066, 2003.

PIRES, A. S. T. **Phase diagram for the two-dimensional quantum anisotropic XY model. Journal of Magnetism and Magnetic Materials, Amsterdam,** v. 260, p. 397399, 2003.

POPINIGIS, F. Seed physiology. Brasilia: AGIPLAN, 289p, 1977.

RABAL, H. J.; BRAGA JÚNIOR, R. A.; TRIVI, M. R.; DAL FABBRO, I. M. The use of lasers in agriculture. In: XXVII CONBEA, Poços de Caldas, 1998.

RABELO, G. F. Evaluation of the application of dynamic speckle in orange quality monitoring. PhD Thesis. Campinas: UNICAMP, 149p, 2000.

RODRIGUES, D.L.; LOPES, H. M.; SILVA, E. R. Soaking, physiological conditioning and effect of sodium hypochlorite on lettuce seed germination. **Revista Trópica - CiênciasAgrárias e Biológicas**, v. 6, n.1, p. 52, 2012.

RODRIGUES, S. **Analysis of Laser Biospeckle in the Differentiation of Live and Dead Tissues in Seeds**. Master's thesis - Federal University of Lavras, 41p, 2003.

RODRIGUES, S. **Stochastic Effect in Dynamic *Speckle***. Thesis (Doctorate) - Universidade Estadual de campinas, 2007.

RODRIGUES, S.; BRAGA JR., R.A.; RABELO, G.F.; FABBRO, I.M.D.; ENES, A.M. Application of dynamic speckle in the analysis of Phaseolusvulgaris L. bean seeds.) **Revista Brasileira de Produtos Agroindustriais**, Campina Grande, v.9, n.2, p.119- 124, 2007.

ROGEZ, H.; BUXANT, R.; MIGNOLET, E.; SOUZA, J.N.S.; SILVA, E.M.; LARONDELLE, Y. Chemical composition of the pulp of three typical Amazonian fruits: araçá-boi (Eugenia stipitata), bacuri (Platoniainsignis),and cupuaçu (Theobromagrandiflorum). **European Food Research Technology**, v.218, p.380-384. 2004.

ROMERO, G. G. **Study and characterisation of speckle patterns that vary over time.**1999. 133 f. Thesis (Doctorate in Physics) - Faculty of Exact Sciences, National University of Salta, Salta, 1999.

SEITZ, K. Cytoplasmic streaming and cyclosis of chloroplasts. In: HAUPT, W.; FELNLEIB, M. E.(ed.). **Encyclopedia of plant physiology**, new ser. Berlin: Springer- Verlag, p.150-169, 1979.

SERT, M.A.; BONATO, C.M.; SOUZA, L.A. Seed germination. In: SOUZA, L.A. (org.) Sementes e plântulas: germinação, estrutura e adaptação: Ponta Grossa: **TODA PALAVRA**, 2009, p.91-118.

SILVA, E.R. **Study of the properties of biospeckle and its applications**. Master's thesis, University of São Paulo. São Paulo, SP, 2007.

SILVA, F. C. **Aplicação do** Speckle **Dinâmico para o Estudo do Processo de Secagem de Pintura.** São Paulo: USP, 2000. 67 p. Master's thesis.

SOUZA FILHO, M.S.M.; LIMA, J.R.; NASSU, R.T.; BORGES, M.F. Physical, chemical and sensory evaluation of native fruit nectars from the North and Northeast of Brazil: Exploratory Study. **Brazilian Journal Food Technology**, v.5, p.139-143, 2002.

SOUZA, E. B; PACHECO, M. V; MATOS, V. P; FERREIRA, R. L. C. Germination of adenantherapavonina L seeds as a function of different temperatures and substrates. **Revista Árvore**, v.31, n.3, p.437-443,2007.

XU, Z, JOENATHAN, C, KHORANA, B. M. Temporal and spatial properties of the timevarying speckles of botanical specimens. **Optical Engineering**, v. 34,n.5, p.1487 1502, 1995.

MIX
Papier aus verantwortungsvollen Quellen
Paper from responsible sources
FSC® C105338

Printed by Books on Demand GmbH, Norderstedt / Germany